CATALOGUE

I0049170

ILLUSTRATED LIST

===== OF =====

BIT BRACES,

Hand and Breast Drills, Chucks,

Pin Vises, Hollow Handle

Tool Sets, Etc.

MANUFACTURED BY

THE JOHN S. FRAY CO.

BRIDGEPORT, CONN., U. S. A.

The Hall & Bill Printing Company
Willimantic, Conn.
1911

FACTORY ON CRESCENT AVENUE
BRIDGEPORT. CONN., U. S. A.

Ball-Bearing Steel Clad Heads.

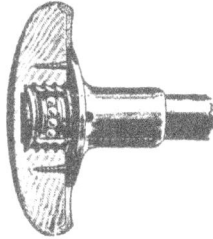

As Used In

This Ball-Bearing device consists of two hardened steel washers, one at the inner end of the iron thimble, the other at the shoulder of the brace crank where milled up to receive it. Between these hardened steel washers the balls revolve, the whole placed within the thimble thus being secured from dirt.

Sectional View of No. 106 Style Brace Socket, Pages 12, 13, 21 and 22.

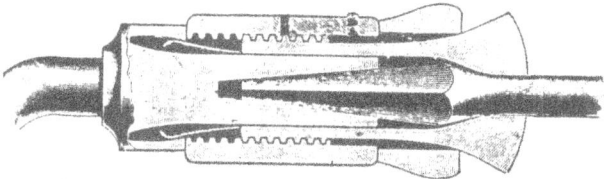

The square taper shank of a bit fits into a corresponding cavity in the chuck or socket and is driven thereby, the forged steel jaws grasping the bit over the shoulder, holding it firmly there.

The inner ends of the jaws are threaded to conform to the thread in the sleeve; and in case of jaws being removed or new ones inserted, care should be taken to see that the outer ends are even with each other before they are drawn in by the turning of the sleeve, (otherwise, they cannot center true). If not even, turn back the sleeve to permit of the outer jaw entering the next thread, pressure being used on the ends to aid in entering the threaded part. The small screw is to prevent the jaws coming out.

Ratchet Braces.

Showing our D or Decagonal Sleeves as furnished when so ordered with our No. 101, 102 and 103 line of braces, price as for regular goods. See pages 6, 8 and 10.

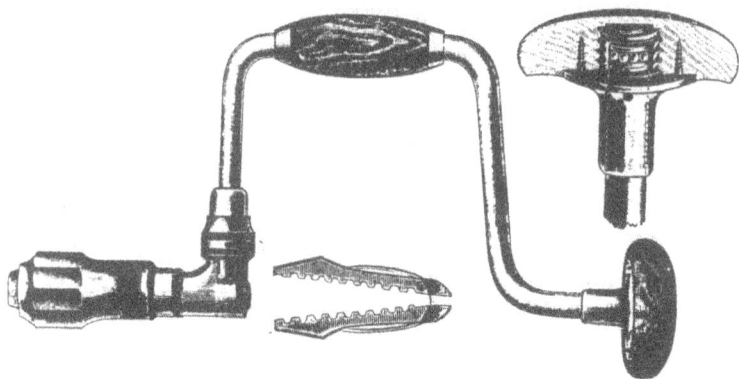

Plain Braces.

Showing our D or Decagonal Sleeves with our No. 210, 310 and 410 line of braces, price as for regular goods. See pages 7, 9 and 11.

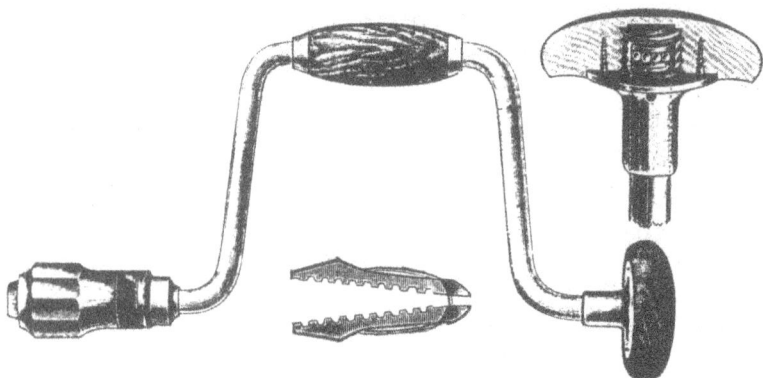

Drill Brace.

Correctly Cut Gears.   Ball Bearing Head.

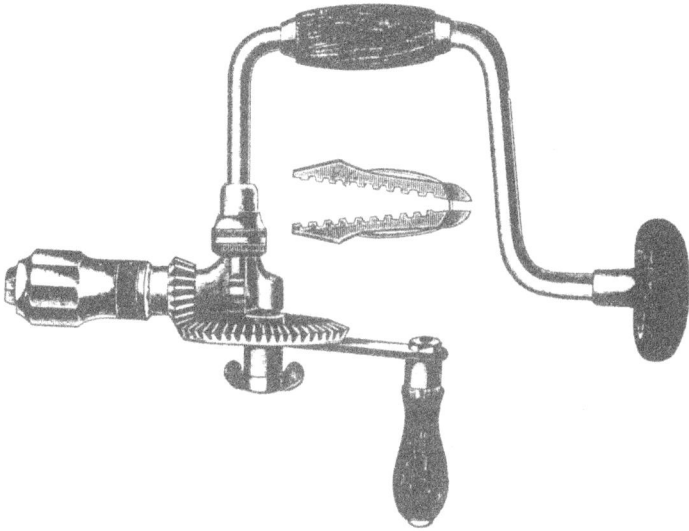

Consisting of our No. 101, best grade 10 inch sweep ratchet brace, to which has been added a breast drill attachment, readily applied or detached.

The brace is furnished with forged steel alligator jaws, which take either round or square shanks.

The action of the ring for operating the pawls differs from our ordinary ratchet braces, in that we turn as far as we can in either direction to raise both pawls, but if one only, turn in the direction for this, until the position of the ring is where it is ready to commence to lift the other. This will be readily understood in using the brace.

These braces are fully guaranteed in every particular, and are furnished with our D or Decagonal Sleeves.

Put up one in a box.   List, $36.00 per dozen.

Ratchet Braces.

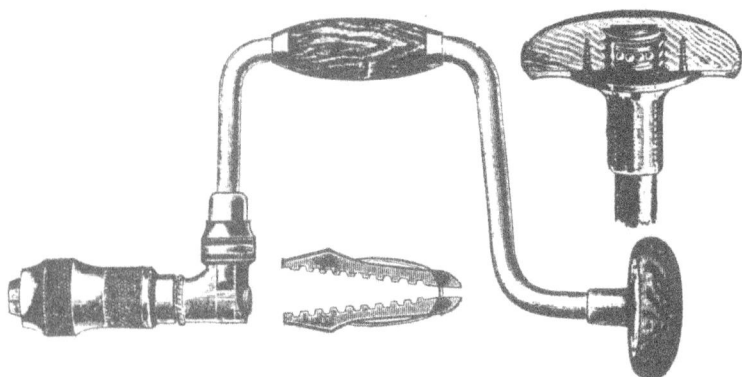

List

									PER DOZ.
No. 61	-	-	-	6 inch	-	-	-	-	$28.00
No. 81	-	-	-	8 inch	-	-	-	-	29.00
No. 101	-	-	-	10 inch	-	-	-	-	32.00
No. 121	-	-	-	12 inch	-	-	-	-	35.00
No. 141	-	-	-	14 inch	-	-	-	-	38.00
No. 161	-	-	-	16 inch	-	-	-	-	43.00

Our Ratchet Braces above illustrated are first class in every respect, and fully guaranteed. The Sweep, Socket, Jaws, Ratchet and Thimble are Steel; Handle and Head are of Cocobolo; all metal parts are Nickel Plated, and fully finished throughout. The peculiar features of these Braces are the Internal Cam Ring to operate the Pawls, and the Spring (of best wire) attached to our Jaws, which gives them perfect automatic action in operating when liberated by the unscrewing of the Sleeve, and in no way interferes with their fitting any shape bit or tool shank.

Decagonal Sleeves (see page 4) in place of the above if preferred.

Put up one-half dozen in a box.

Ball Bearing Heads

Our heaviest braces of this class as No. 121, 141 and 161 have special heavy frames, ratchet and socket, thus adapting them for the heaviest work.

Sleeve Braces.

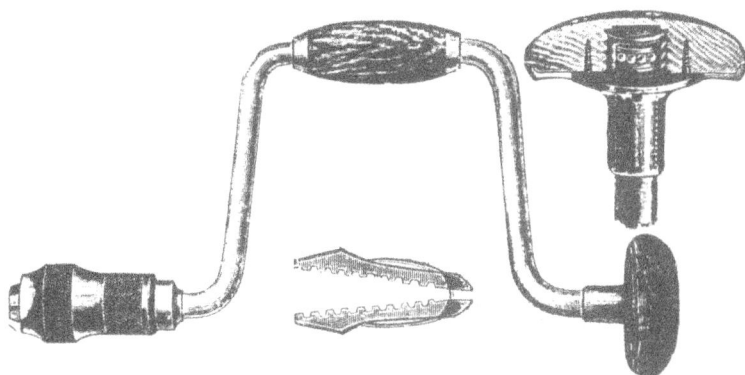

List

								PER DOZ.
No. 206	-	-	-	6 inch sweep	-	-	-	$21.00
No. 208	-	-	-	8 inch sweep	-	-	-	24.00
No. 210	-	-	-	10 inch sweep	-	-	-	27.00
No. 212	-	-	-	12 inch sweep	-	-	-	30.00
No. 214	-	-	-	14 inch sweep	-	-	-	33.00

These goods are the same grade as our best Ratchet Braces; with Socket and Jaws of similar construction; Sweep, Socket, Jaws and Thimble Steel; all metal parts Nickel Plated; Cocobolo Head and Handle.

Decagonal Sleeves (see page 4) in place of the above if preferred.

Put up one-half dozen in a box.

Ball Bearing Heads.

Ratchet Braces.

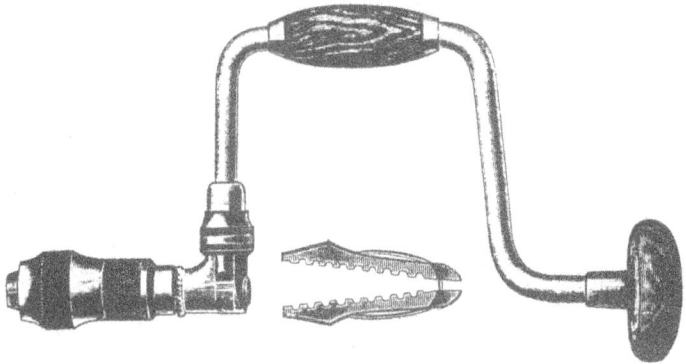

List

										PER DOZ
No. 62	-	-	-	-	6 inch	-	-	-	-	$23.00
No. 82	-	-	-	-	8 inch	-	-	-	-	24.00
No. 102	-	-	-	-	10 inch	-	-	-	-	27.00
No. 122	-	-	-	-	12 inch	-	-	-	-	30.00
No. 142	-	-	-	-	14 inch	-	-	-	-	33.00

These Constitute our Medium-Finished, Nickel-Plated Ratchet Braces; Sweep, Socket and Jaws are Steel; Cocobolo Head and Handle; parts made with same care as No. 61 series, but less highly finished. Heads are not ball bearing.

Our Decagonal Sleeves (see page 4) in place of the above if so ordered.

Put up one-half dozen in a box.

Sleeve Braces.

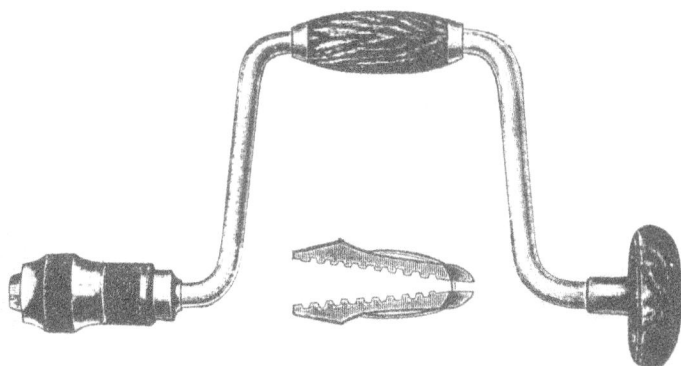

List

								PER DOZ.
No. 306	-	-	-	6 inch sweep	-	-	-	$12.00
No. 308	-	-	-	8 inch sweep	-	-	-	14.00
No. 310	-	-	-	10 inch sweep	-	-	-	16.00
No. 312	-	-	-	12 inch sweep	-	-	-	18.00
No. 314	-	-	-	14 inch sweep	-	-	-	20.00

These Braces have Socket and Jaws like our 206 series; are Nickel Plated, with Cocobolo Head and Handle, and form a medium between our best and the plain-polished braces. Heads are not ball bearing.

Decagonal Sleeves (see page 4) if so ordered.

Put up one-half dozen in a box.

Ratchet Braces.

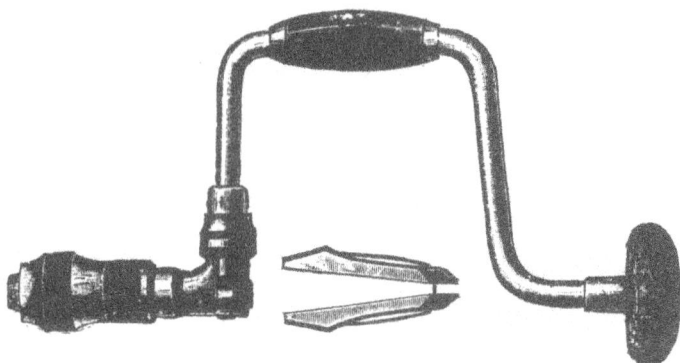

List

										PER DOZ.
No. 83	-	-	-	-	8 inch	-	-	-	-	$18.00
No. 103	-	-	-	-	10 inch	-	-	-	-	19.00
No. 123	-	-	-	-	12 inch	-	-	-	-	21.00
No. 143	-	-	-	-	14 inch	-	-	-	-	23.00

These are our best grade of plain polished Ratchet Braces. Metal parts same as No. 62 Series (page 8), including Forged Steel Jaws and Socket, but not Nickel Plated. Head and handle Ebonized Hard wood, or, if so ordered, Mahogany Stain finish.

Alligator style Jaws in place of plain, furnished if preferred, with the above braces.

Sleeve Braces.

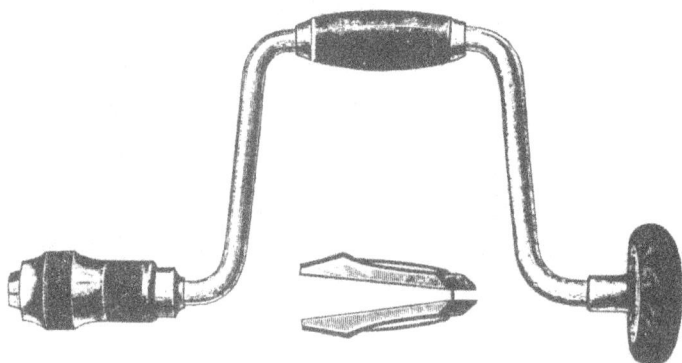

List

								Per Doz.
No. 408	-	-	-	8 inch sweep	-	-	-	$ 9.00
No. 410	-	-	-	10 inch sweep	-	-	-	10.00
No. 412	-	-	-	12 inch sweep	-	-	-	11.00
No. 414	-	-	-	14 inch sweep	-	-	-	12.00

Plain Polished Sleeve Braces, not Nickel Plated, but with Jaws and Socket, same style as our best and medium (pages 7 to 9): Head and Handle Ebonized Hard Wood; (or, if so ordered, Mahogany Stain finish); a good Brace for service.

Alligator style Jaws in place of plain, furnished if preferred, with the above braces.

Ratchet Braces.

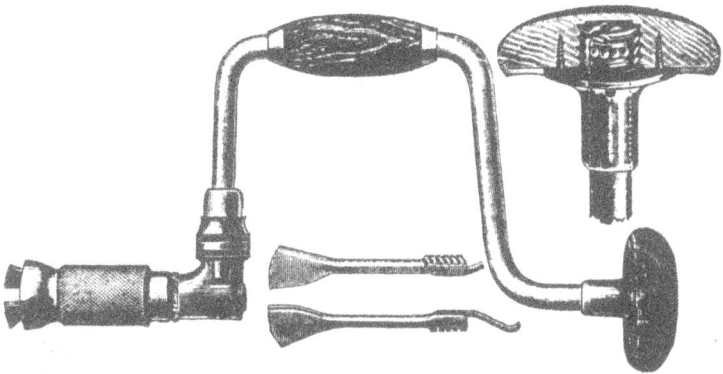

List

										Per Doz.
No. 66	-	-	-	-	6 inch	-	-	-	-	$28.00
No. 86	-	-	-	-	8 inch	-	-	-	-	29.00
No. 106	-	-	-	-	10 inch	-	-	-	-	32.00
No. 126	-	-	-	-	12 inch	-	-	-	-	35.00
No. 146	-	-	-	-	14 inch	-	-	-	-	38.00
No. 166	-	-	-	-	16 inch	-	-	-	-	43.00

These braces differ from others in our catalogue (see page 3) in that the square shank of the bit fits into a socket and is driven thereby; the jaws, holding firmly by the round part of bit shank beyond the square, prevent its coming out. These have

Ball Bearing Heads.

Crank, Sleeve, Jaws and Thimbles are Steel. Head and Handle of Cocobolo Wood. Metal parts Nickel Plated. The hardened jaws open automatically, as the sleeve is unscrewed, until fully open, when a stop-screw arrests the further motion of both sleeve and jaw.

Round shank drills may be used in these braces.

Nos. 126, 146 and 166 have special heavy ratchet, frame and socket, thus adapting them to heavy use.

Sleeve Braces.

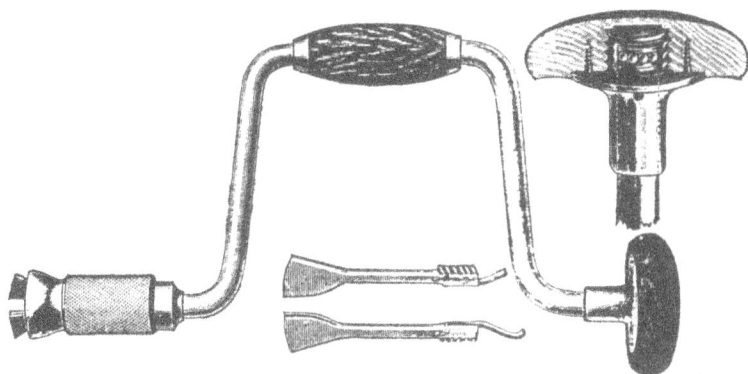

List

									Per Doz.
No. 606	-	-	-	-	6 inch	-	-	-	-	$21.00
No. 608	-	-	-	-	8 inch	-	-	-	-	22.00
No. 610	-	-	-	-	10 inch	-	-	-	-	24.00
No. 612	-	-	-	-	12 inch	-	-	-	-	27.00
No. 614	-	-	-	-	14 inch	-	-	-	-	30.00

These are our plain Sleeve Braces, with Socket, Sleeve and Jaws, same as our Ratchet Braces Nos. 66 to 166, having our

Ball Bearing Heads

We make but one grade—the best—of both Plain and Ratchet, in this line of Braces.

Put up one-half dozen in a box.

Ratchet Braces.

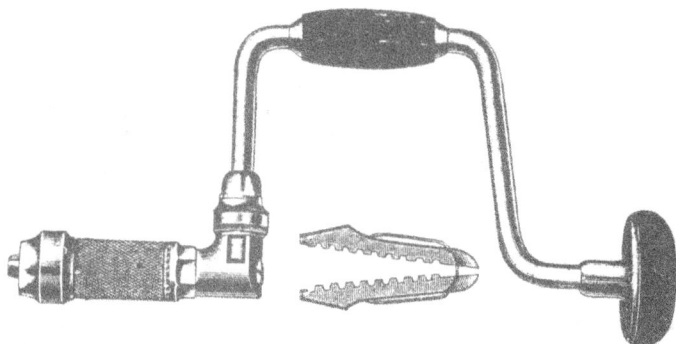

										PER DOZ.
No. 109	-	-	-	-	10 inch	-	-	-	-	$52.00
No. 129	-	-	-	-	12 inch	-	-	-	-	55.00
No. 149	-	-	-	-	14 inch	-	-	-	-	60.00
No. 169	-	-	-	-	16 inch	-	-	-	-	66.00

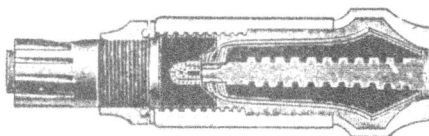

SECTIONAL VIEW BALL BEARING CHUCK

The chuck or bit holding device, as see illustration, consists principally of a socket, sleeve, binder, ball-retaining ring and the hardened steel jaws; the ratchet is formed at one end of the socket, while the outer end is made heavier than usual to resist the extra strain caused by the ball bearing at the end of the sleeve, which enables the operator to exert greater power in closing the jaws on to a bit or drill; these jaws are specially formed to take round shank drills from 1-16 to about ½ inch, as also square shank bits.

At the end of the sleeve is a steel washer, next to which are shown the balls, which, with the recessed ring, threaded to prevent any possibility of shifting laterally, and kept from turning by a small screw, forms the ball bearing sleeve device.

Should it become necessary at any time to renew the jaws, first remove the socket from the ratchet frame; then take out the small screw in the recessed ball-retaining ring, which then can be unscrewed from the socket, (care being taken not to lose the balls therein). This will leave the sleeve free to be taken from the binder and socket, when both jaws and binder may be removed.

It will be noted that we show a wire connection between the binder and spring of the jaws; while this is not essential, its object is to give a positive return motion to the jaws.

Spofford Brace.

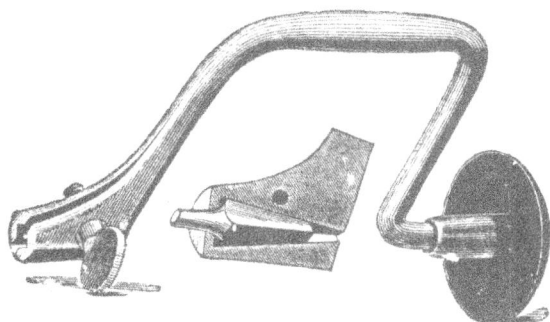

List

								PER DOZ.
No. 7	-	-	-	7 inch sweep	-	-	-	$16.00
No. 8	-	-	-	8 inch sweep	-	-	-	19.00
No. 10	-	-	-	10 inch sweep	-	-	-	22.00
No. 12	-	-	-	12 inch sweep	-	-	-	25.00
No. 14	-	-	-	14 inch sweep	-	-	-	28.00

Our Spofford Braces are all furnished with thumb-screws forged from the best Bessemer Steel, so as to give all possible strength thereto.

Spofford Braces.

Cocobolo Wood Head and Handle. Nickel Plated

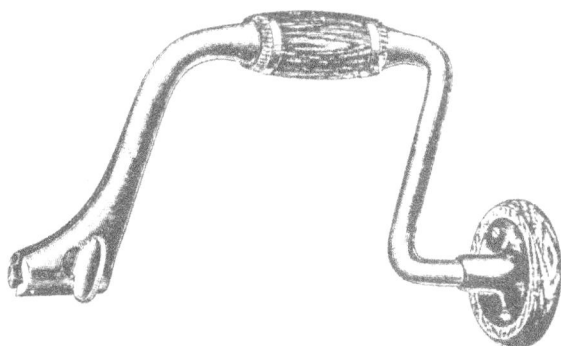

No. 107	-	-	-	7 inch sweep	-	-	-	$21.00
No. 108	-	-	-	8 inch sweep	-	-	-	24.00
No. 110	-	-	-	10 inch sweep	-	-	-	27.00
No. 112	-	-	-	12 inch sweep	-	-	-	30.00
No. 114	-	-	-	14 inch sweep	-	-	-	33.00
No. 117	-	-	-	17 inch sweep	-	-	-	36.00

Sleeve Braces.

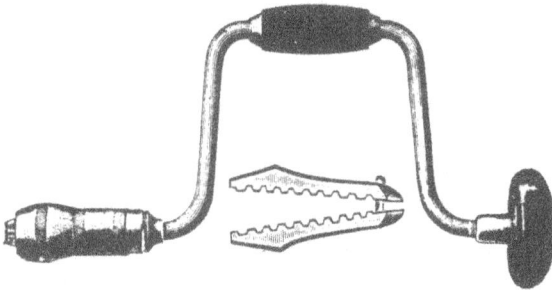

No. 508	-	-	-	8 inch sweep	-	-	-
No. 510	-	-	-	10 inch sweep	-	-	-
No. 512	-	-	-	12 inch sweep	-	-	-

Ratchet Braces.

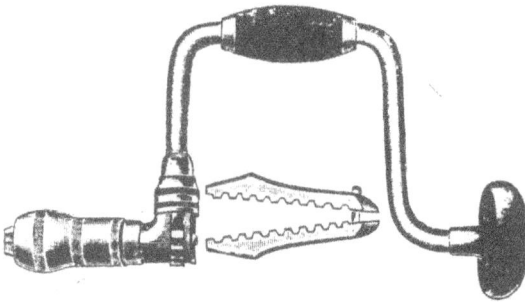

No. 85	-	-	-	8 inch sweep	-	-	-
No. 105	-	-	-	10 inch sweep	-	-	-
No. 125	-	-	-	12 inch sweep	-	-	-

The above Braces grade next to our No. 83 Ratchet (page 10) and 408 Plain (page 11) but are not quite so heavy; and in place of our Forged Steel Spring Jaws with Jaws and Socket milled to fit each other, we dispense with milling and use only Plain Jaws, as shown in the cut. We furnish these Braces with Sleeves, as on page 17, for our export trade.

One-half dozen in a box.

Sleeve Brace.

No. 508-A	-	-	8 inch sweep	-	-	Nickel Plated.
No. 510-A	-	-	10 inch sweep	-	-	Nickel Plated.
No. 512-A	-	-	12 inch sweep	-	-	Nickel Plated.
No. 508-B	8 inch sweep, Nickel Plated, Cocobolo head and handle.
No. 510-B	10 inch sweep, Nickel Plated, Cocobolo head and handle.
No. 512-B	12 inch sweep, Nickel Plated, Cocobolo head and handle.

Ratchet Brace.

No. 85-A	-	-	8 inch sweep	-	-	Nickel Plated.
No. 105-A	-	-	10 inch sweep	-	-	Nickel Plated.
No. 125-A	-	-	12 inch sweep	-	-	Nickel Plated.
No. 85-B	8 inch sweep, Nickel Plated, Cocobolo head and handle.
No. 105-B	10 inch sweep, Nickel Plated, Cocobolo head and handle.
No. 125-B	12 inch sweep, Nickel Plated, Cocobolo head and handle.

These Braces are the same as on page 16, except that they are Nickel Plated and finished accordingly, the letter A or B in addition to the number indicating this finish. PLAIN SLEEVES, as on page 16, will be sent with these Braces unless where the trade calls for the above style.

One-half dozen in a box.

Sleeve Brace.

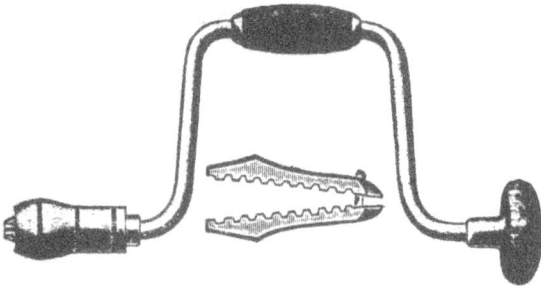

No. 0108	-	-	-	8 inch sweep	-	-	-
No. 0110	-	-	-	10 inch sweep	-	-	-
No. 0112	-	-	-	12 inch sweep	-	-	-

Ratchet Brace.

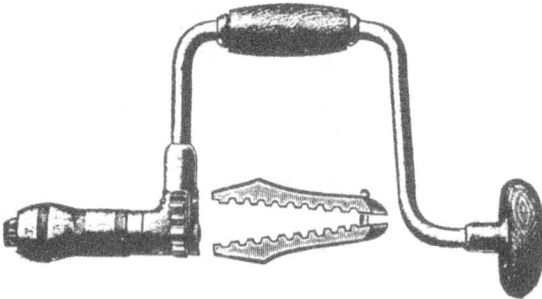

No. 1085	-	-	-	8 inch sweep	-	-	-
No. 1105	-	-	-	10 inch sweep	-	-	-
No. 1125	-	-	-	12 inch sweep	-	-	-

The above Braces have Steel Sweep, Polished Bright, Head and Handle Hard Wood, Stained Cherry, Chuck and Jaws, etc., as shown by cut. The same grade goods as on page 19, No. 08, 085, etc., but having heads with iron quill, giving all metal bearing thereto.

One-half dozen in a box.

Sleeve Braces.

No. 08	-	-	-	8 inch sweep	-	-	-
No. 010	-	-	-	10 inch sweep	-	-	-
No. 012	-	-	-	12 inch sweep	-	-	-

Ratchet Braces.

No. 085	-	-	8 inch sweep Ratchet	-	-
No. 0105	-	-	10 inch sweep Ratchet	-	-
No. 0125	-	-	12 inch sweep Ratchet	-	-

These are our Farmers' Braces, Steel Sweep, Polished Bright, Head and Handle Common Hard Wood, Red Stained, Chuck and Jaws, etc., as shown by cut.

One-half dozen in a box.

Improved Angle Boring Bit Stock.

Nickel Plated.

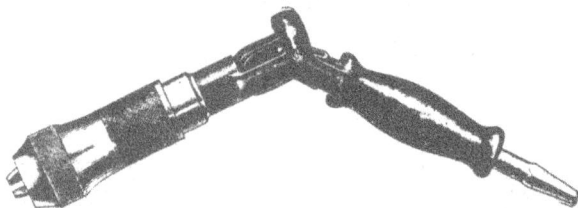

This tool is protected at the universal joint by a portion of the frame, which forms a band passing over the joint. The frame being in one piece, is light, and at the same time strong and durable. The Chuck is the same as on our Braces on pages 6 and 7, having our Patent Spring Jaws.

Put up in paper boxes. One-half dozen in a box. $24.00 per dozen.

Spofford Double Crank or Whimble Brace.

For Millwrights, Ship Carpenters, Etc.

Nickel Plated.   Handles, Cocobolo.

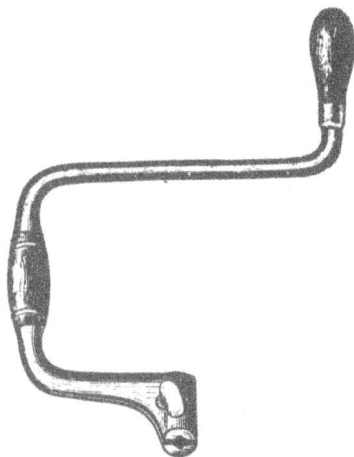

| No. 10 | - | - | - | 10 inch sweep | - | - | - | $33.00 |
| No. 12 | - | - | - | 12 inch sweep | - | - | - | 36.00 |

Corner Bit Brace.

Nickel Plated.

Head and Handles of Coco-bolo Wood.

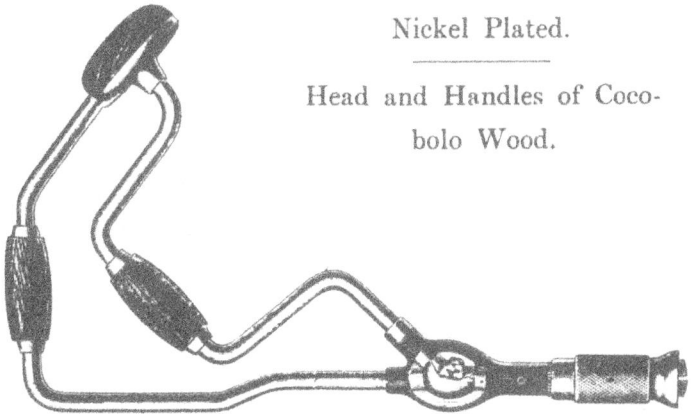

No. 80, 8 inch sweep   –   –   ..   List $33.00 per Dozen

No. 100, 10 inch sweep   –   –   –   List 36.00 per Dozen

The Improved Corner Brace, as here illustrated, will be found in these days of special requirements in the line of electric appliances, plumbing, etc., to materially aid in making the necessary openings for such. The Ratchet Brace now an indispensable tool, is not rapid in its use as such, taking as much or more time in its backward movement as in its forward or cutting course; unlike this, the corner brace is driven constantly in either direction with complete revolutions, in cramped positions, corners, etc., as with an ordinary brace unobstructed.

These are made in two sizes, viz: Eight and ten inch sweep, and are in all respects, first-class goods, being finished in the best style. The gears are of wrought steel, and owing to the shape in which the teeth are cut, are almost frictionless.

The Cocobolo Head is reinforced or strengthened by an iron-clad thimble or quill; one of the three screws fastening this thimble to the wood head goes through the steel bow-shaped portion of the frame where inserted into the head, thus securing head, bow and thimble together.

The style of chuck adopted is the same as used on Fray's No. 106 Ratchet Brace, now so well known.

In the above illustration the guard-plate covering the gears is shown partially removed, thereby showing a section of the gears. In addition to keeping the gears free from dirt, these plates prevent possible injury in handling the brace.

Put up two in a box.

No. 70 Corner Ratchet Bit Brace.

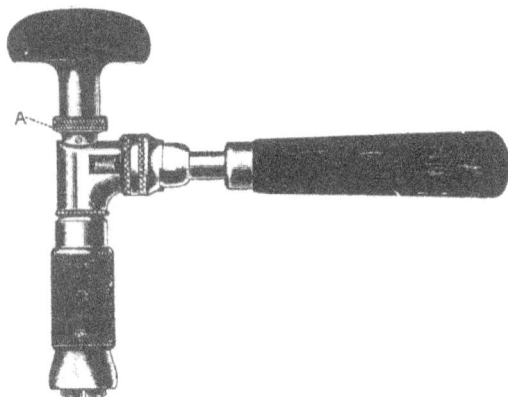

Designed for use in cramped positions not admitting of ordinary braces.

The chuck is the same as used on our No. 106 line of braces, see page 3.

The entire tool, both in material and workmanship, is first-class. Metal parts are nickel plated; head and handle of cocobolo wood.

Between the head and ratchet frame is placed a knurled collar, marked "A" on cut: the use of this is to assist in starting a bit or tool previous to its entering the wood or other material sufficiently so that the friction thereof shall hold it from turning back whilst the handle is turned to take a fresh hold on the ratchet; the operator, with the finger and thumb of the hand supporting the head, bearing slightly on this collar, holds the chuck, preventing its turning only as driven by the ratchet.

Put up two in a box.

List per Dozen $18.00.

Extension Bit Holders.

Nickel Plated.

Barber Style Chuck.

No. 1

This is so well known that the cut alone is sufficient to show its style. Socket and jaws are wrought steel.

No. 2

Chuck same as furnished in Fray's line of braces Nos. 66 to 166. Jaws are of forged steel.

No. 3

⅜ Inch Extension, consisting of a socket formed on the end of the steel extension rod, forged steel jaws, and sleeves of drawn steel tubing, hence the entire tool is of steel. These goods have been recently improved, by which greater strength and efficiency have been given them.

No. 4

¾ Inch Extension is, as our ⅜, made entirely of steel, while the additional size gives increased strength beyond that of the ⅜ or No. 3, proportionate to the increased size.

Nos. 1, 2, and 3 are put up one-half dozen in a box. No. 4 is put up one-third dozen in a box.

All of these are made in 12, 16 and 20 inch lengths.

We furnish, to order, 24 and 30 inch lengths at an advance to cover extra cost.

Price per dozen of Nos. 1, 2 and 3, $18.00 List.

Price per dozen of No. 4, $21.00 List.

Bit Brace Parts and Prices.

Catalogue No. 25.

When ordering any part state number of brace for which part is wanted.

NO.		PAGE	PRICE
1	Drill Brace Frame . .	5	$ .30
2	Closed Ratchet Frame .	6, 8 & 12	.25
3	Open Ratchet Frame .	10	.25
4	Open Ratchet Frame .	18	.20
5	Sleeve	5	.25
6	Sleeve	6, 8 & 10	.25
7	Sleeve	16	.20
8	Sleeve	12	.25
9	Socket	6 & 8	.25
10	Socket	12	.30
11	Socket	10 & 16	.20
12	Pawls, per pair . . .	6 & 12	.25
13	Pawls, per pair . . .	10	.25
14	Pawls, per pair . . .	16	.10
15	Pawls, per pair . . .	18 & 19	.10
16	Jaws, per pair . . .	6 & 8	.25
17	Jaws, per pair . . .	16, 18, &c.	.10
18	Jaws, per pair . . .	12 & 13	.25
19	Head, Maple	19	.08
20	Head, Maple, and Thimble	10, 16, &c.	.15
21	Head, Cocobolo, and Thimble	6, 12. &c.	.30
22	Head, Cocobolo, and Wide Thimble . .	6, 12, &c.	.30
23	Ratchet Wheel . .	10, 16, &c.	.15
24	Socket Screw	6, 8 & 12	.10
25	Socket Screw Washer .		
26	Jaws and Binder . . .	14	.30

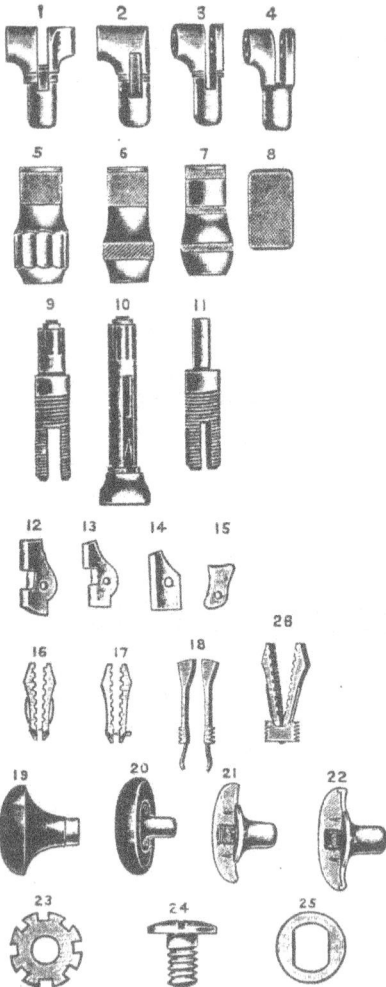

Hand Drill No. 3.

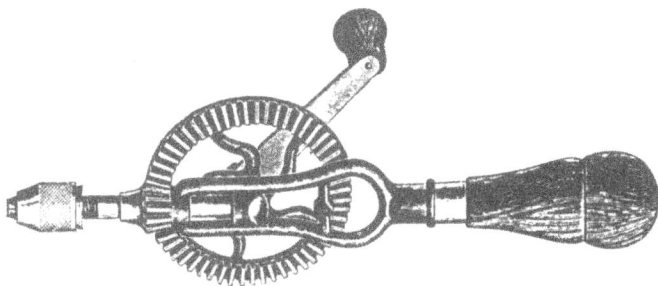

This Hand Drill has Malleable Iron Frame, Spindle and Pinion of Steel, correctly cut Gears, 3-Jaw Chuck, taking from 0 to 5-32. Made of the best material and workmanship. Handles of Cocobolo. With the above we furnish 6 drills, from 1-16 to 9-64 inch. Length over all, 10½ inches; weight, 1¼ pounds. List $15.00 per dozen.

Hand Drill No. 4

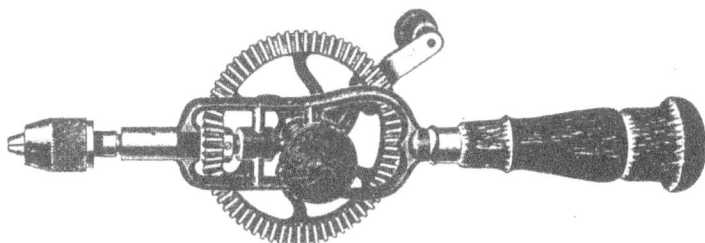

Frame of Malleable Iron, with Spindle and Pinion of Steel, as No. 3, but larger, thus forming a medium between our No. 3 Hand Drill and our Breast Drills.

Chuck is three-jawed taking from 0 to ¼ inch, and is furnished, as No. 3, with six drills, placed in the handle. Metal parts nickel plated. Handles, Cocobolo; length, 12½ inches; weight 1¾ pounds. List $24.00 per dozen.

The above drills put up one in a box.

Breast Drill No. 1.

With Level.

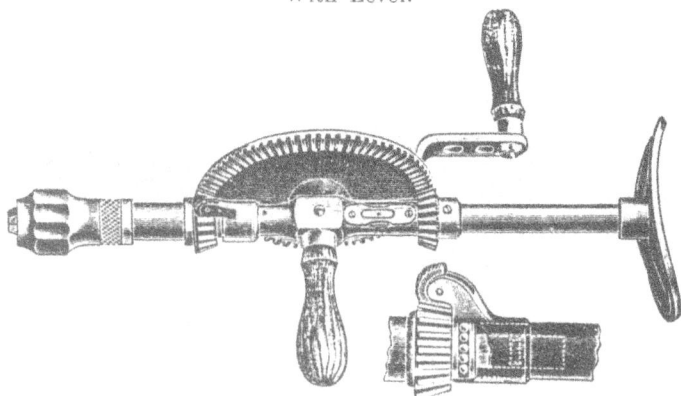

Our No. 1 Breast Drill, as above illustrated, equals in size any on the market; is double geared, all gears cut, frame wrought steel, with alligator forged steel jaws, hardened.

It will be seen we have a ball bearing shoulder to take up the end thrust, as also a special locking device by which the socket is securely held from turning by a latch instantly thrown in or out by the thumb of the left hand whilst holding the breast drill in position for use, thus materially assisting in tightening or releasing drills. All goods finished in the best manner and fully nickel plated.

Breast Drill No. 5

With Level.

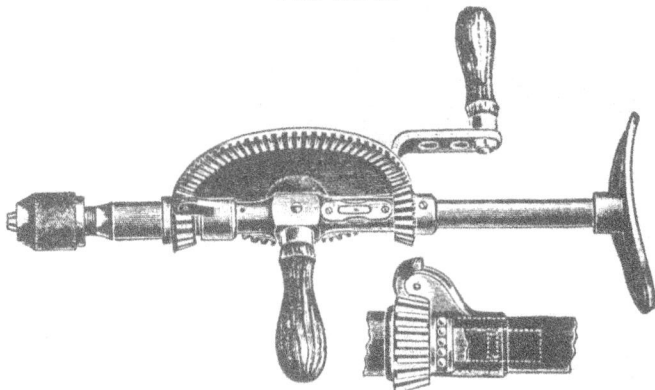

This Drill does not differ in construction from our No. 1, except that in place of the brace chuck, it is furnished with a three-jaw chuck, taking from 0 to 7-16 inch. Chuck has tool steel hardened jaws, same as our chucks, page 41.

List of each $48.00 per dozen.

DISCOUNT_____

Breast Drill No. 2.

This Drill is furnished with change of speed; that as shown, being three revolutions of the chuck to one of the crank, the other, for heavier work, at about equal speeds for both chuck and crank.

As with our No. 1, the socket has a ball-bearing shoulder to take up the end thrust, and a special locking device by which the socket is securely held in place when needed to tighten or change drills. Furnished with Alligator Forged Steel Jaws. Gears are cut. Frame, Malleable Iron. List $30.00 per dozen.

Breast Drill No. 6.

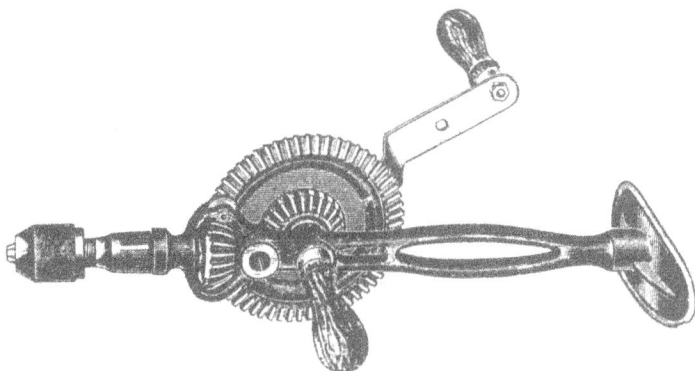

Construction and parts are precisely as our No. 2, but in place of the brace chuck, as there shown, is provided with a three-jaw chuck, which takes from 0 to 7 16 inch round shanks. Jaws are tool steel, hardened. See page 41.

Put up one in a box. List $39.00 per dozen.

Breast Drill No. 7.

With Level.

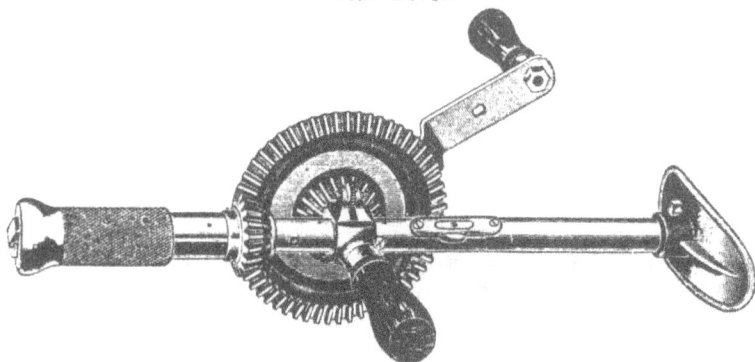

This Breast Drill, as with our Nos. 2 and 6, has change of speed, being from about even turn of crank and chuck, to three of chuck to one of crank. The change of speed is accomplished by loosening the screw which forms a part of the supporting handle sufficiently to allow the larger gear to tilt back out of gear; then slide the moveable gear stud from the one position to the other, and tighten; the point of this screw fits into countersunk depressions, which, in addition to other stops at each end, permits of no misplacing. A locking device, to hold the chuck from turning when tightening or releasing drills, is provided. The crank is pierced to allow of change of position of the drive handle from eight up to eleven inch sweep of crank. The chuck is practically the same as our No. 106 line of braces, but much heavier. It holds securely from 1-32 to 7-16 inch round and square shank tools. One in a box. List $38.00 per dozen.

Breast Drill No. 8.

With Level.

As No. 7, illustrated above, but with Barber Style Chuck and Alligator Steel Jaws. List $36.00 per dozen.

Breast Drill No. 9.

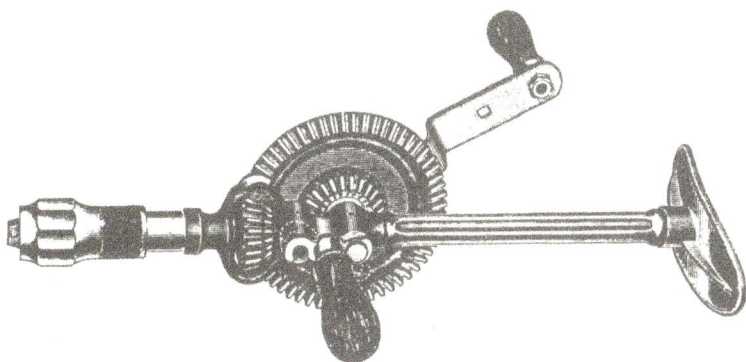

Frame of Malleable Iron, strong and at the same time, light.

Chuck, Barber style, with steel alligator jaws, to take both round and square shank tools.

Ball-bearing thrust. Cut Gears. Handles of Black Walnut.

To operate the latch which holds the large gear spindle in place in changing speed, as also the latch which holds the chuck from turning in changing drills, it is not necessary to remove the hand from the supporting handle, as both latches are readily controlled by the thumb and forefinger, the hand still retaining its hold on the handle. Each operation is almost instantaneous.

The gear spindle latch is held in place by an automatic stop on the inside of the latch, which while offering little resistance in shifting, is sufficient to hold when in place.

We hesitate not to say that this Breast Drill, though intended to supply a demand for a comparatively cheap tool, will be found, on trial, to be a thoroughly reliable and desirable article.

Finish:—Chuck, crank, steel pinions and shifting latch, polished.

Frame, breast plate and large wheel, japanned.

List $24.00 per dozen.

Grip Handle for Breast Drills.

This may be attached to any of our Breast Drills.

Price per dozen, $3.00.

If Breast Drills are ordered with grip handle instead of breast plate, price same as with breast plate.

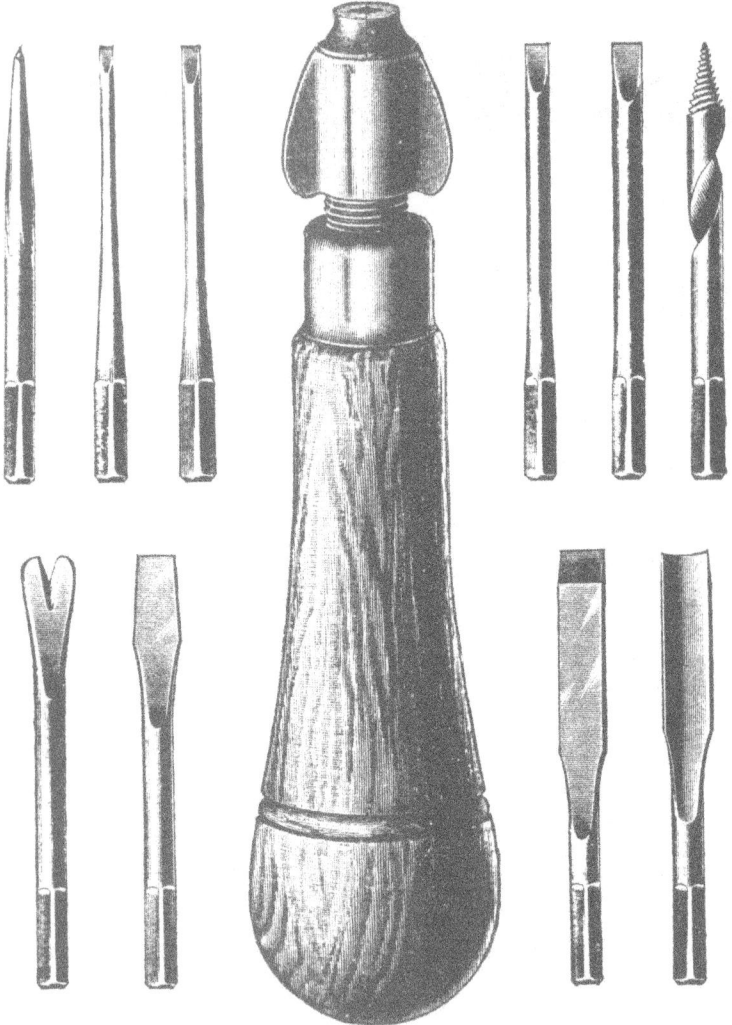

Fray's Patent
Hollow Handle Tool Sets.

No. 1 Full Size.

Fray's Patent
Hollow Handle Tool Sets.

No. 1–C. Full Size.

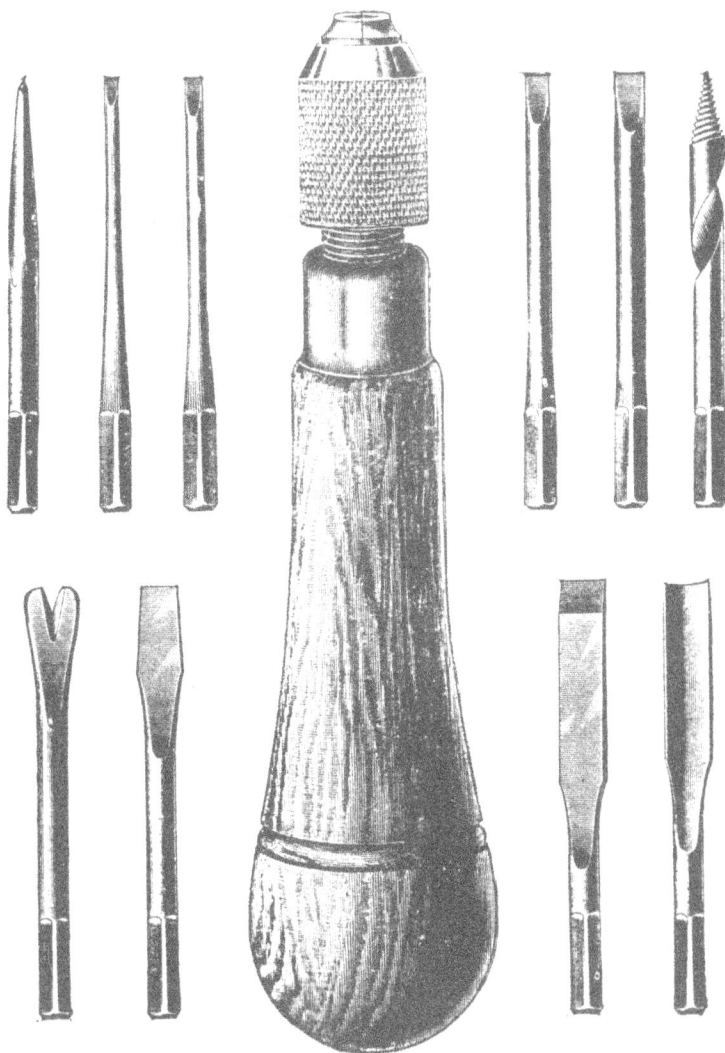

Fray's Patent Awl and Tool Sets.

No. 1.

These Handles and Tools are first class in every respect. The Handle is Cocobolo Wood. The Jaws, Clamping-Nut and Ferrule are Nickel Plated.

Cut Half Size.

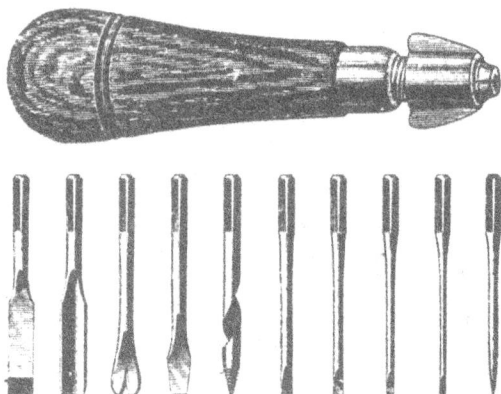

The Tools Consist of

Chisel, Tack-Puller, Gouge, Gimlet, Screw-Driver, Reamer, Scratch-Awl and Four Brad-Awls of Different Sizes.

The Shanks of the tools are three-sixteenths of an inch and are squared largest at the end, to prevent their drawing out of the socket when in use. They are made of Cast Steel, properly tempered.

Put up one-half dozen in a box.

List $12.00 per dozen.

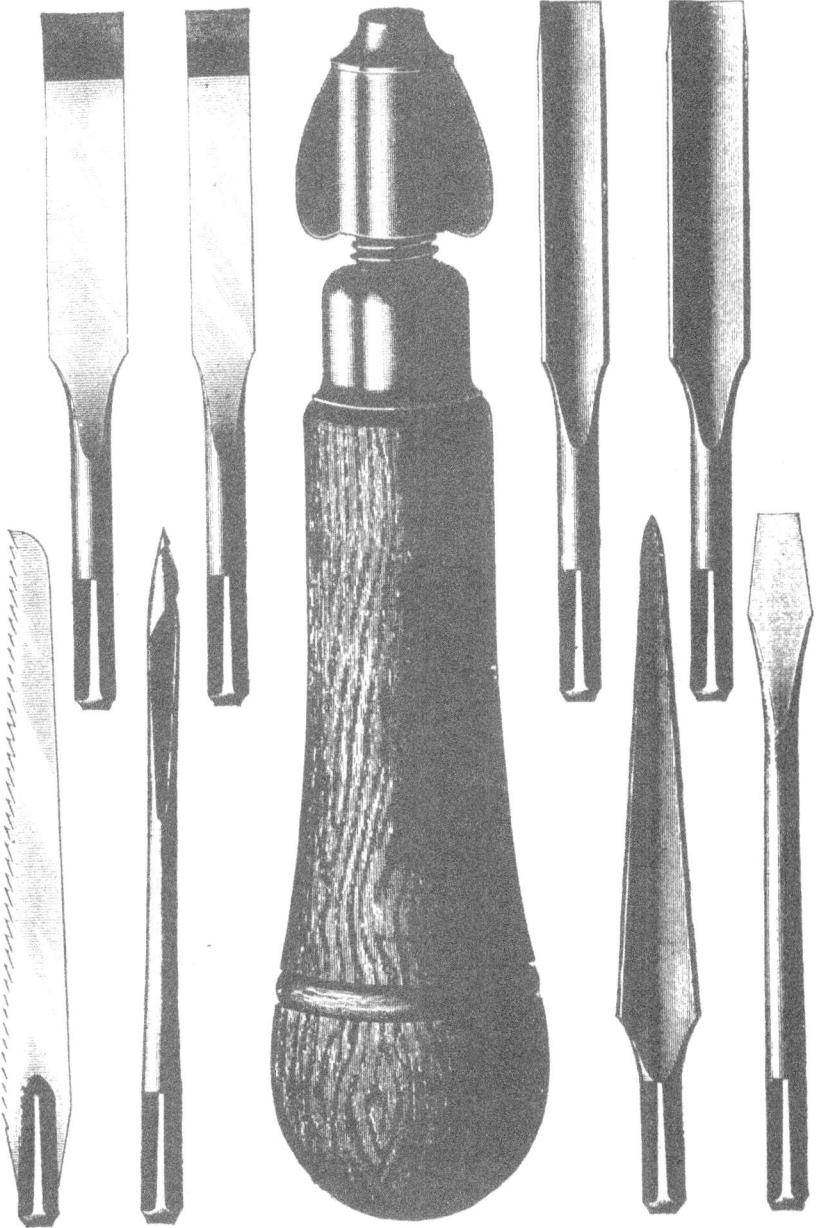

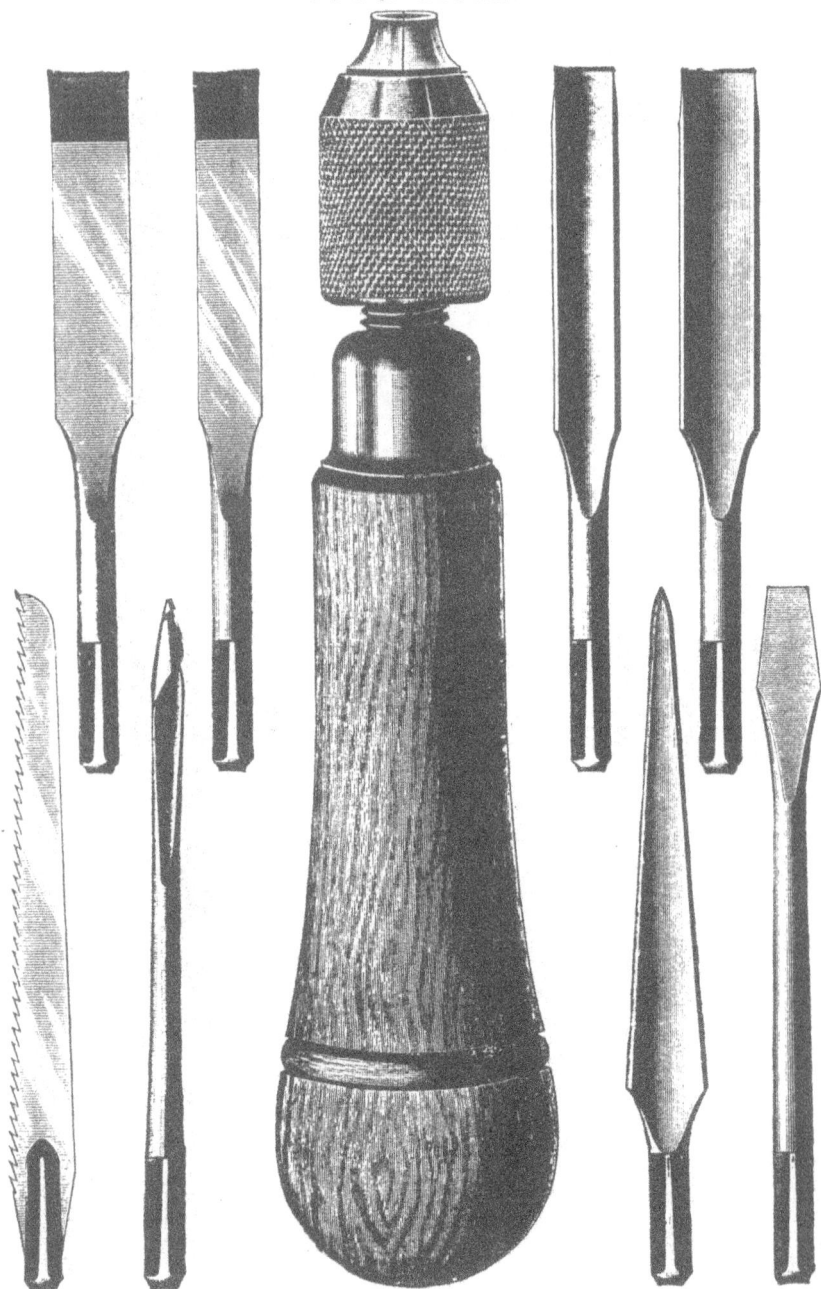

Fray's Hollow Handle Tool Sets.

No. 2-A and 2-C.

Cuts Full Size as Shown on Preceding Pages.

The No. 2 Handle is the same in construction as No. 1, only larger. The tools consist of two sizes each of *Chisels* and *Gouges*, one *Screw Driver*, one taper *Reamer*, one *Gimlet Bit*, and one *Saw*. All are made for use, and are sufficiently large for general work, etc. The saw blades are made for us by one of our best saw manufacturers. The tools are of regular tool steel, hardened in oil, and carefully finished.

Combination No. 2-B.

Cut Half Size.

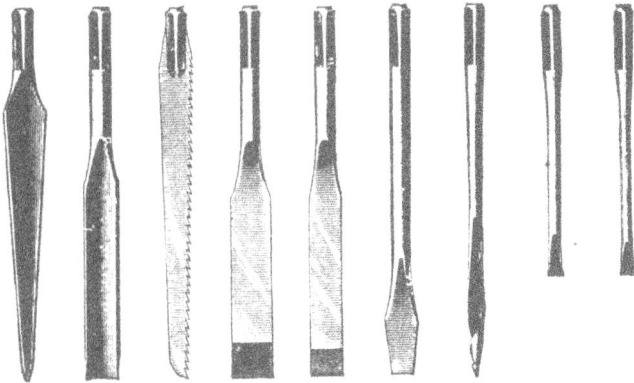

We have for our No. 2 Handles a combination of tools, consisting of two large Shank Brad-Awls of suitable size, in place of the Gouge, making nine tools instead of eight, as above. This we call "No. 2-B." Price of each the same. Saws 6¾ inches long furnished in place of the 4½ inch, if so ordered, at 25c per dozen net extra. List $18.00 per dozen.

No. 2C with eleven tools, including file and long saw, at $1.00 per dozen, extra.

No. 6 Apple Wood Handle.

With eight tools, one each Square Reamer, Short Saw, Gouge, Screw Driver, Gimlet, Brad-Awl, and two Chisels. Handle and tools same size as No. 2 on page 33. List $16.00 per dozen. Put up one-half dozen in a box.

Fray's No. 4 Hollow Handle Set.

Cocobolo Handle.

Cut Half Size.

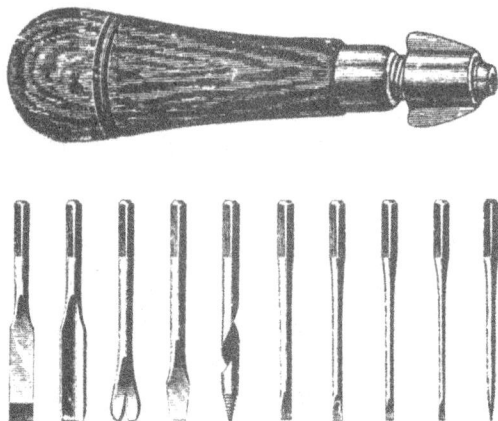

This Hollow Handle Set has been brought out by us to meet the demand for a cheaper grade of this class of goods than our regular No. 1 Hollow Handle Set, which, with our No. 2 A, B and C, as also No. 3 Sets, we propose to keep fully up to the present standards. But realizing that perhaps in the majority of cases the surroundings are such as to call for a tool to do the work, with less regard to nicety of the finish, etc., found in the more expensive class of goods, we have no hesitancy in recommending this as meeting such a demand.

Number and arrangement of Tools as above.

Put up one-half dozen in a box.

List $9.00 per dozen.

Fray's Hollow Handle Set No. 5.

Handle of Native Hard Wood; Maple, Etc. Combination of Tools
as No. 4. List $7.00 per dozen.

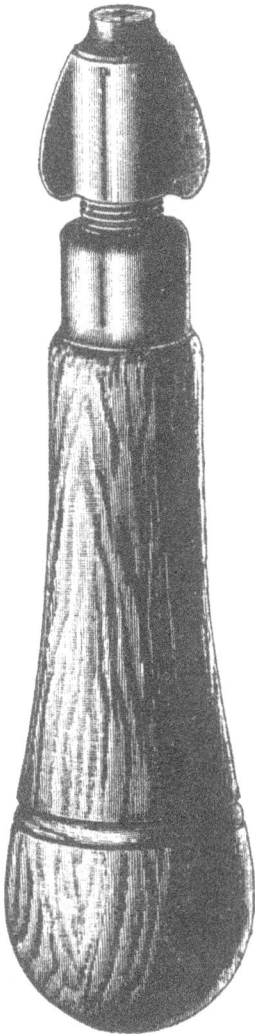

Tools Half Size.

Put up ½ Dozen in a Box.

No. 5 Handle. Full Size.

Fray's Hollow Handle
Screw Driver Set.
No. 3.

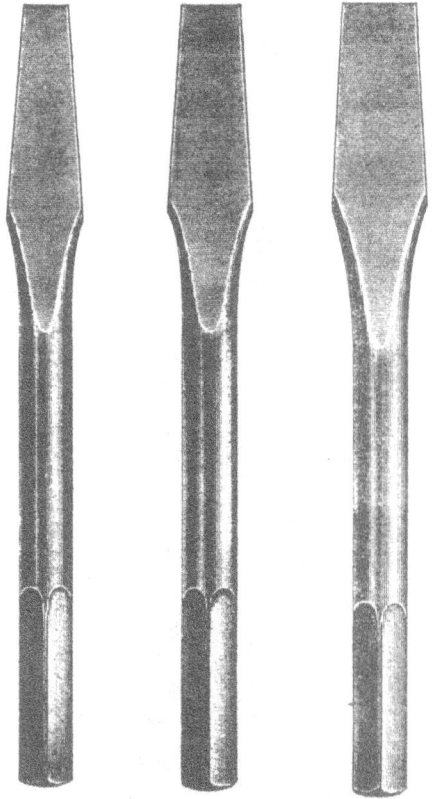

Cut is Full Size.

Price $12.00 per dozen.

These goods have been recently improved by a brass nickel plated cap, which screws into a metal thimble, let into the hollow end of the handle containing the three assorted screw driver bits.

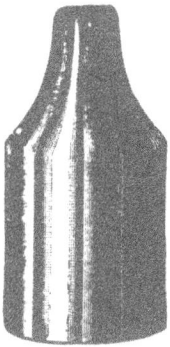

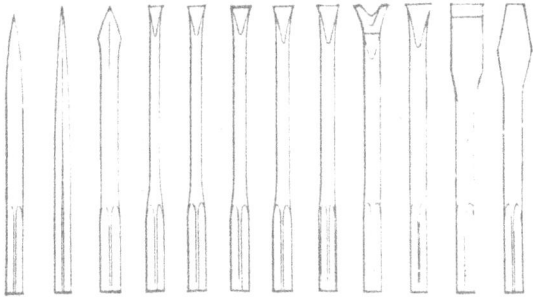

Combination Haft.

This illustration shows these combination Handles and Tools full size, making a convenient tool, especially adapted for light jobbing and repairing.

No danger of the handle splitting, for it is solid wood.

After removing the metal cap, the tools all stand in view for selection.

Special pains are taken to make these tools first class in all respects.

No. 7, Cocobolo Handle,   Per Dozen $12.co
No. 8, Maplewood Handle, ·   Per Dozen 11.co

Pin Vise—Patent Chuck.

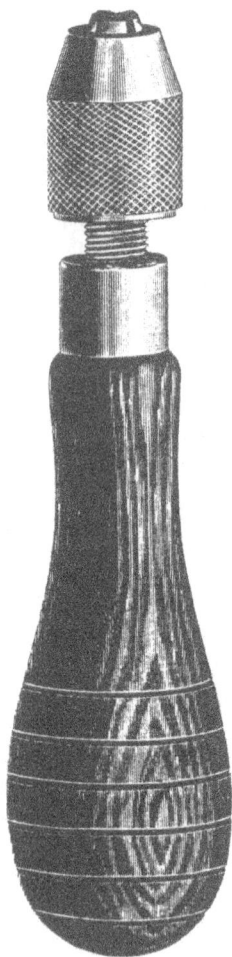

The above consists of a Patent 3-Jaw Chuck, same size and style as we furnish with our Hand Drill, No. 3. The handle is Cocobolo and drilled to admit of wire or other articles passing through.

The cut is full size. We do not hesitate to say this forms one of the most complete and useful tools of its class. Packed one in a box.

List $15.00 per dozen.

Patent Three Jaw Chucks.

The most simple and effective of this class of tools.   Durable, Accurate.

The three hardened steel jaws are held apart by spiral springs, which, as the Chuck is loosened, draw them back automatically.

	DIAM.	LENGTH	PRICE EACH
No. 1, Capacity 0 to 5-32 in.	Shank, 7-16 in.	2 in.	$1.50
No. 2, Capacity 0 to ¼ in.	Shank, ½ in.	2 ½ in.	3.00
No. 3, Capacity 0 to ½ in.	Shank, 9-16 in.	2 ½ in.	4.00

If furnished with Morse Taper Shank, No. 1 or 2, 50 cents extra each.

Packed one in a box.

www.ingramcontent.com/pod-product-compliance
Lightning Source LLC
Chambersburg PA
CBHW030535210326
41597CB00014B/1161